Time to Let Go of Some of Those Cherished Physics and Astronomy Dogmas?

Third Edition

John Winfrey

Copyright © 2016 John Winfrey

All rights reserved.

ISBN-13: 978-1540537065

ISBN-10: 1540537064

Cover Photo: Copyright purchased from
http://www.123rf.com/profile_bradcalkins

Table of Contents

CHAPTER 1: THE HUMAN BRAIN AFFECTS RESEARCH 5
The Researcher Influences the Research 5

Paradigms 5

Postulates 6

Empiricism v Rationalism 8

Empiricism's Failings 10

CHAPTER 2: CRACKS IN THE CURRENT PARADIGMS 14
Dark Matter 14

Dark Energy 15

Constants or Laws that May Vary with Space or Time 16

Thermodynamics 16

Special Relativity 19

On Singularities 19

CHAPTER 3: INTERPRETATION OF EXPERIMENTAL MEASUREMENTS CAN BE A SLIPPERY SLOPE 21
The Chain of Verified Principles versus the Chain of Ideas in Teaching Physics 21

The Absolute versus the Relative 22

The "Relative" Is Best Demonstrated By The Doppler Effect For Sound. 25

CHAPTER 4: LIGHT, THE BEARER OF QUESTIONABLE TRUTH 27
    Electric Field Subtleties Hidden from Students 27

    Basis of Second Postulate of Relativity? 30

    More Ways Light Can Deceive the Experimentalist 31

    Words Can Cause Confusion 31

CHAPTER 5: SPECIAL RELATIVITY IN DETAIL 35
    The Michelson-Morley Experiment 35

    What about the Singularity? 42

    More Ways Light Deceives 44

    Einsteinian Relativity 49

CHAPTER 6   DARK MATTER 54
    The Historical Setting 54

    My Perspective as a Physicist 57

    What Kind of Mechanism Might We Look For? 59

    My Theory: The Change in Physics within the Disc 60

    What is this "New Force"? 62

CHAPTER 7: CONCLUSIONS 65

CHAPTER 1: THE HUMAN BRAIN AFFECTS RESEARCH

The Researcher Influences the Research

Research and discovery are NOT independent of the nature of the researcher, much as we would like to believe so!

The human brain is capable of great leaps in insight, but also has limitations. Such limitations are finite memory capacity, language, ego, and incomplete and inaccurate access to truth because of brain and sensory organs processing structures[1].

Hence, to compensate, some investigative restrictions are employed:

Paradigms

A paradigm is

> A set of assumptions, concepts, values, and practices that constitutes a way of viewing reality for the community that shares them, especially in an intellectual discipline.[2]

The upside of a paradigm is that it limits (accommodating our brain's limited size) the amount of information involved in any research sub discipline. The downside is that if the solution one is seeking is outside a given paradigm, the solution cannot be found.

Another downside: there is a book[3], required reading 50 years ago when I was an undergraduate, which describes the rather large resistance displayed within research communities to a paradigm change (heresy). That resistance is in contrast to the common, smug mission statement of openness to new truth.

---

[1] The human brain does not encode information in a one-one way, and our sensory systems are nonlinear and limited in range.
[2] The Free Dictionary.
[3] "The Structure of Scientific Revolutions". Thomas Kuhn, 1962.

Postulates

A postulate is

> Something assumed without proof as being self-evident or generally accepted, especially when used as a basis for an argument.[4]

The use of postulates comes from the Rationalism approach to thinking: what the human mind can reason or imagine to be true is true regardless of experience or authority.[5]

A postulate is meant as a starting point, an island of security. The intent is to eventually understand the basis of the postulate, but that sometimes does not happen.

Examples:

The Cosmological Principle:

> The universe is uniform [always the same, as in character or degree; unvarying], homogenous [consisting of parts that are the same; uniform in structure or composition], and isotropic [identical in all directions], and therefore appears the same from any position.[6]
>
> *****
>
> The cosmological principle is usually stated formally as 'Viewed on a sufficiently large scale, the properties of the universe are the same for all observers.' This amounts to the strongly philosophical statement that the part of the universe which we can see is a fair sample, and that the same physical laws apply

---

[4] The Free Dictionary.
[5] Rationalism has its roots in the Christian Church. Scriptures that say, "God thought and it was." This ability became ascribed to humans, who in earlier centuries were thought to be only slightly lower in rank than God.
[6] Wikipedia.

throughout. In essence, this in a sense says that the universe is knowable and is playing fair with scientists.[7]

Postulates of Relativity:[8]

1. First postulate (Principle of Relativity)

The laws of physics are the same in all inertial frames of reference [coordinate systems moving at constant speed with respect to each other].

2. Second postulate (Invariance of the speed of light)

The speed of light in free space has the same value c in all inertial frames of reference.

*****

This two-postulate basis for special relativity is the one historically used by Einstein, and it remains the starting point today. As Einstein himself later acknowledged, the derivation of the Lorentz transformation tacitly makes use of some additional assumptions, including spatial homogeneity, isotropy, and memorylessness. Also Hermann Minkowski implicitly used both postulates when he introduced the Minkowski space formulation, even though he showed that c can be seen as a space-time constant.

---

[7] William C. Keel (2007). The Road to Galaxy Formation (2nd Ed.). Springer-Praxis. p. 2.
[8] Wikipedia.

Empiricism v Rationalism

How do we establish truth? Two approaches:[9]

> **Rationalism** is the view that 'regards reason as the chief source and test of knowledge' or 'any view appealing to reason as a source of knowledge or justification'. More formally, rationalism is defined as a methodology or a theory 'in which the criterion of the truth is not sensory but intellectual and deductive'....
>
> Rationalists believed that reality has an intrinsically logical structure. Because of this, the rationalists argued that certain truths exist and that the intellect can directly grasp these truths. That is to say, rationalists asserted that certain rational principles exist in logic, mathematics, ethics, and metaphysics that are so fundamentally true that denying them causes one to fall into contradiction. The rationalists had such a high confidence in reason that empirical proof and physical evidence were regarded as unnecessary to ascertain certain truths – in other words, 'there are significant ways in which our concepts and knowledge are gained independently of sense experience'.
>
> **Empiricism** in the philosophy of science emphasizes evidence, especially as discovered in experiments. It is a fundamental part of the scientific method that all hypotheses and theories must be tested against observations of the natural world rather than resting solely on *a priori* reasoning, intuition, or revelation.
>
> Empiricism, often used by natural scientists, says that 'knowledge is based on experience' and that

---

[9] Wikipedia.

'knowledge is tentative and probabilistic, subject to continued revision and falsification.'

Mathematics is a rational endeavor. Physics is an empirical endeavor. Because the human mind unfettered is very productive, there is far more Mathematics than Physics. See Venn diagram below:

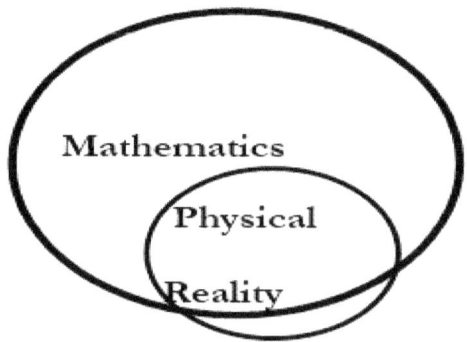

Figure 1-1. The domains of Physics and Mathematics.

Physics employs only Mathematics that describes physical reality. And yes, there is now reality for which there is no Mathematics: Chaotic events.

In my opinion, many contemporary scientists have left the heritage of Newton, and are no longer explaining reality *as it is*. They are far too often describing nature *as it must be* as seen in their minds and Mathematics.

This is a dangerous, and an historically error-prone approach.

I tell my Physics students, "Nature rules the Mathematics in this course, not the reverse!" Why? Novice problem solvers often bring into play arbitrary Mathematics in problems they don't fully understand in order to produce "a" solution.

Empiricism's Failings

On the other hand ….

Empiricism was and is a reaction to rampant Rationalism.

Empiricism basically says, "What we can *measure* is truth!"

> Counter-example: Copernicus stated that the sun is the relatively motionless body in the solar system, and the Earth is orbiting about the sun.
>
> The Church's argument was that if the Earth were moving, then humans would feel that motion in their sensory organs.[10]
>
> The answer to this dilemma came in the 1900's with Einstein. Einstein pointed out that there was no absolute reference frame for describing motion, and there was no "Physics" experiment that one could perform to detect absolute motion. All motion is relative. If there is no "Physics", there is nothing for "Biology" to build a speedometer in humans.
>
> When you tell the Officer you didn't know how fast you were going, that is the truth; you couldn't detect it physiologically!

However, as will be revealed in a later chapters, designing a valid and definitive measurement"

> o Can be surprisingly difficult
> o Sometimes is impossible[11].

From Newton until Einstein, empiricism went VERY well. In fact, other disciplines tried to adopt this approach with mixed results.

However, with the beginning of the 1900's, the scientific community

---

[10] This is because the Church believed humans were only slightly less gifted than God, and of course He would be able to detect a moving Earth.

[11] Example: One cannot observe or measure a particular star's light if it passed the Earth 5,000 years ago.

began to measure things that they didn't like or expect, or that were counter-intuitive.

The hard-liners (many still living today) insisted that empiricism was the end-all way to go. "What we can measure is THE truth!"[12]

Well, if your measurement device is flawed, sloppy or misinterpreted, or if there are areas of "forbidden knowledge", this won't work.

The human ego seems to cause some researchers to selectively forget these crucial loop holes in empiricism!

A good example of fooling ourselves into believing we "understand" is Quantum Mechanics. This topic gives students much heartburn and not all instructors are skilled enough to get these students through *this* topic with integrity. I have read author after educated author misinterpret Quantum Mechanics on multiple fronts.

At the heart of Quantum Mechanics is the Uncertainty Principle: one CANNOT measure simultaneously the position and the momentum (for the casual reader, the speed) of a piece of matter. For the macroscopic world, that imprecision is completely negligible.

I tell introductory students: when one is investigating the smallest pieces of matter, we don't have a "tool" that is smaller than these small objects of interest which can probe them without disturbing them.[13]

---

[12] There is an old story. Moses, when he was coming down from speaking with God, was carrying three gold tablets with commandments. He got tired half way down, and left the third tablet behind, hidden in bushes. The 11th commandment on the third tablet was, "Thou shall not be stupid, my little children."

[13] This is called "the observer effect". It is (at this time) a practical limit. There are examples of the Uncertainty Principle at work that show the principle has deeper roots.

(Incidentally, light has simultaneously BOTH properties of a particle and of a wave.[14] And the particle property of light, the photon, has the same properties and behaviors as any of the other elemental particles. It is NOT insubstantial in the world of the very small. If I *were* God, I would have done this very differently....)

The example I give students is:

> Suppose you are in a dark, empty room which contains a ping pong ball on the floor with zero speed (at rest). Your Task: find and measure where the ball is with ultimate accuracy.
>
> Well, you strike out probing with your fingers (or as a good scientist, with your Calipers) to tentatively locate the ball. Once you touch it, you know approximately where it "was", but you have imparted motion to the ball that is significant. You have, now forever in its future, imparted an unknown speed.
>
> It's like squeezing down from both sides on a wet watermelon seed: the more you squeeze to precisely locate it, the more violently it squirts out of your fingers.

The very process of measuring in the world of the elemental particle changes the particle's future motion.

Bad explanations say, "The process of *observing* small pieces of matter changes their future". The novice will think of their experience in the macroscopic world (where the uncertainty principle is trivially met): I can look at my car, and afterward it will still be the same. That is because "looking" or "observing" means shining light (or a stream of air molecules or of water) at my car and seeing the probing medium reflect off it.

If you were to "observe" your car by sending a stream of other cars at it (because you don't have anything smaller available) and

---

[14] Again, something we don't "like" because in the macroscopic world, waves and particles are distinct.

observing reflections, indeed your car would have a different future.

This is "forbidden knowledge" in Physics! It is a LIMIT to measurement built into the Laws and fabric of Physics.

Other limits to measurement will be explored in later chapters.

CHAPTER 2: CRACKS IN THE CURRENT PARADIGMS

Let's take a look at some of the indicators of a pending paradigm shift.

Dark Matter

Astronomers have found that galaxy stars do not revolve the same way planets do in our solar system (inside planets travelling faster, Kepplerian Dynamics).

In fact, in the disc region, stars rotate at the same tangential speed regardless of radius. See Figure 6-2 in Chapter 6.

The Astronomer's fix: Dark Matter which does not emit light as "regular matter" does, and which is plentiful and which lies between stars in a galactic disk, and accounts for what *must be* the dynamics for the disc (not necessarily *what is* the dynamics).

> **Dark matter** is an unidentified type of matter comprising approximately 27% of the mass and energy in the observable universe that is not accounted for by dark energy, baryonic matter (ordinary matter), and neutrinos The name refers to the fact that it does not emit or interact with electromagnetic radiation, such as light, and is thus invisible to the entire electromagnetic spectrum. Although dark matter has not been directly observed, its existence and properties are inferred from its gravitational effects such as the motions of visible matter, gravitational lensing, its influence on the universe's large-scale structure, and its effects in the cosmic microwave background. Dark matter is transparent to electromagnetic radiation and/or is so dense and small that it fails to absorb or emit enough

radiation to be detectable with current imaging technology.

Estimates of masses for galaxies and larger structures via dynamical and general relativistic means are much greater than those based on the mass of the visible "luminous" matter.[15]

Dark Energy

Similarly, dark energy is employed to fix the *apparent* finding that the farthest stars from Earth are <u>accelerating</u> outwards, which is inconsistent with the <u>single explosion</u> dynamics of the Big Bang Theory.

> In physical cosmology and astronomy, **dark energy** is an unknown form of energy which is hypothesized to permeate all of space, tending to accelerate the expansion of the universe. Dark energy is the most accepted hypothesis to explain the observations since the 1990's indicating that the universe is expanding at an accelerating rate.
>
> Assuming that the standard model of cosmology is correct, the best current measurements indicate that dark energy contributes 68.3% of the total energy in the present-day observable universe.[16]

Dark Matter and Dark Energy hide all these otherwise unexplained effects within constructs that, by definition, are unlikely to be detected and unlikely to be confirmed!

This approach is not philosophically honest nor is it proper empiricism.

Already, there has been a partial recant: apparently the acceleration is

---

[15] Wikipedia
[16] Wikipedia

less than estimated and the amount of Dark Matter in the universe is smaller.[17] Again, this is a measurement issue. Astronomers calibrate their experimental methods on near stars where the information is more certain, and then use the Cosmological Principle to extend the results to distant stars.

In this case, the near calibration supernovae were *intrinsically* less intense than the distant supernovae studied. The Cosmological Principle was incorrect and the "tool" flawed.

Constants or Laws that May Vary with Space or Time

I always tell my students, "I am teaching Physics HERE and NOW. I cannot with certainty guarantee that these Principles may not vary with space and time as you move away from Earth."

I remember as a graduate student hearing about experiments that showed the gravitational constant, G, is drooping slightly with time. This seemed, at the time, to be consistent with Universal expansion. The community said this experiment needed to be verified by other researchers.

Jump forward to a current update, and in a new document[18] researchers are finding an alternative explanation of Supernovae Type 1a behavior (as opposed to Universal accelerating expansion), that the results may be consistent with temporal changes in G.

There are, however, counter arguments that this temporal droop in G, although real, is too small to account for the theory above.

Thermodynamics

The Laws of Thermodynamics were historically devised to explain Earth-bound observations, where all systems are relatively close to

---

[17] "Accelerating Universe? Not so Fast". Science Daily, April 11 2015.
[18] E. Garcıa–Berro, Y. Kubyshin, and P. Loren–Aguilart & J. Isern, "The variation of the gravitational constant inferred from the Hubble diagram of Type Ia supernova". Retrieved from: https://arxiv.org/pdf/gr-qc/0512164v2.pdf.

equilibrium[19] and dominated by friction (which is an irreversible and pervasive process). The primary application is heat engines. This is what is currently presented in introductory Physics texts. Cosmologists know only this result, and build their theories upon this premise, leading to a Universe moving toward disorder.

However, there is a more recent interpretation of the forward direction of time[20], taking into account systems that are far from equilibrium, i.e. have regions with vastly different energy levels. In fact, Cosmology may be a better fit with this perspective.

> If a system is moved away from thermodynamic equilibrium by the application of a [large] gradient of energy [coined *exergy*], an attractor[21] for the system can emerge for the system to organize in a way that reduces the effect of the applied gradient [more efficiently]. I.E., if the system becomes more organized, allowing more receptors for the excess energy to flow into, the system will self-organize to more quickly and reduce the energy gradient.
>
> Some systems in physics that self-organize are:
>
> - Lasers: coherent light, self-organization of many atoms
> - Nonlinear optics: coherent light, self-focusing, generation of harmonics, coherent Raman and Brillouin scattering, etc.
> - Fluid dynamics, gas dynamics: cloud streets, convection instability, Taylor-Couette flow, roll

---

[19] Near equilibrium, there are no pockets of very high or very low energies (temperatures).
[20] There is a Nobel Prize for this work. However, that discovery was made by a Chemist and Physicists and Astronomers don't regularly follow Chemistry discoveries.
[21] An attractor is a set of numerical values toward which a system tends to evolve, for a wide variety of starting conditions of the system. System values that get close enough to the attractor values remain close even if slightly disturbed.

patterns, hexagonal patterns (Bénard cells), weak turbulences, defects, etc.
- Gas discharges: patterns of molecular densities under the impact of electromagnetic fields.
- Plasma physics: density and velocity patterns of partly or fully ionized atoms and electrons in (partly self-organized) electromagnetic fields, instabilities.
- Semi conductors: patterns of electron and hole densities and currents, Gunn-effect, current filaments.
- Astrophysics: formation and structure of planets, stars, galaxies, big bang, voids, etc.
- Meteorology: climatology, cloud formations, cyclones, etc.
- Geophysics/Geodynamics: inner and surface structure of the earth, geodynamo
- Hadron plasmas: formation of hadron plasmas in high energy collisions of hadrons.
- Self-sustained oscillations: can be found in many of the above mentioned fields.
- Radio-engineering and other sources of coherent electromagnet fields: magnetron, clystrons, etc.

In these systems, forward-in-time is indicated by more organized systems.[22]

Self-organizing systems don't violate the Laws of Thermodynamics. The higher order system is localized, and the compensating disorder is discharged elsewhere.

---

[22] Retrieved from: http://www.scholarpedia.org/article/Self-organization#Physics

Special Relativity

Special Relativity deals with what observers in inertial frames of reference (imbedded in coordinate systems travelling at constant speed relative to each other) are *predicted to observe*[23] in the other coordinate system.

SR theory as presented in text books is based upon the Michelson-Morley experiment[24] which says the speed of light is measured in EVERY coordinate system to be the constant c = 3 x 10^8 m/s.

The hallmark of the theory is a correction factor γ to most Classical Mechanics measurements

$$\gamma = \frac{1}{\sqrt{1 - \frac{v^2}{c^2}}} \qquad (2\text{-}1)$$

where v is the relative velocity between coordinate systems.

There is a problem with this formulation when the relative velocity between observers becomes v = c. This is a theory with a singularity.[25]

The Physics fix: Disallow an observer (matter) to travel at the speed of light in ANY coordinate system of measurement.

On Singularities

Let's take a side trip. I recall when, as an undergraduate, we were studying the driven mechanical oscillator. If one just employs the standard restoring force and a driving force, one finds that the

---

[23] Note: Observation is not equal to absolute truth. The observer at rest relative to their coordinate system's event is the most solid, objective truth.
[24] Which we will prove invalid in a later chapter.
[25] Wikipedia: In mathematics, a singularity is in general a point at which a given mathematical object is not defined.... For example, the function f (x) = 1/x on the real line has a singularity at x = 0, where it seems to "explode" to ±∞ and is not defined.

oscillator position becomes:

$$x = \frac{A}{\omega^2 - \omega_0^2} \sin(\omega t + \phi). \qquad (2\text{-}2)$$

where ω_0 is the resonant frequency of the oscillator and ω is the frequency of the driving force.

My professor asked the class to explain the mathematical case $\omega = \omega_0$. When we were silent, he said:

"Mathematics will call this case undefined. *Nature will never allow an undefined, you have just made an initial assumption about a small parameter that may be ignored, which is incorrect in the end result. When you correct that assumption, the undefined will go away.*"

Continuing with the problem, as soon as even a small amount of friction is added

$$f = bv$$

the undefined indeed goes away:

$$x = \frac{A}{\sqrt{(\omega^2 - \omega_0^2)^2 + 4b^2\omega^2}} \sin(\omega t + \phi). \qquad (2\text{-}3)$$

There are two more cases where an undefined was removed: Landau damping[26] and Renormalized QED[27].

---

[26] Landau, L. "On the vibration of the electronic plasma". *JETP* 16 (1946), 574. English translation in *J. Phys. (USSR)* 10 (1946), 25.
[27] N. N. Bogoliubov, D. V. Shirkov (1959): *The Theory of Quantized Fields*. New York, Interscience.

CHAPTER 3: INTERPRETATION OF EXPERIMENTAL MEASUREMENTS CAN BE A SLIPPERY SLOPE

Over the past 50 years, there has been a blurring of the difference between what a meter "reads" in an experiment, and what the reality it is designed to measure "is". These days, meters seem to control Physics rather than Physics controls meters.

That blurring has found its way into classrooms, texts, and the science professional's as well as the general public's understanding of experimental science.

The Chain of Verified Principles versus the Chain of Ideas in Teaching Physics

Physics has been described as Swiss cheese: There is a contiguous piece of cheese, big holes, and no obvious place to start describing the entire structure. Text book writers have a daunting task!

Especially in regard to Special Relativity, the chain of verified evidence begins with Maxwell's Equations, which are in the second half of most introductory courses.

So with Newtonian Physics (the first half of introductory courses), one has the choice of:

- Telling the hard truth up front and losing most students the first week (I know of 1 text that does this)
- Teaching Newtonian Physics first, and then revealing later in the course that you told a half-truth (a few texts)
- In either term, introducing Special Relativity using an old calculation and experiment which is wrong and the interpretation wrong; fix Newtonian Physics somewhere in the text layout; and for people who only take introductory Physics never straighten this out. For Physics majors, maybe NEVER straighten this out, since Newtonian Physics and

Electrodynamics are separate courses, upper division and graduate school.

I propose to tell the story of Special Relativity based upon a staircase of verified Physics steps, wherever each Step lives in the cheese.

Again, the introductory text book writers probably can't do this, which is irksome!

The Absolute versus the Relative

Let's begin with some examples, and use only sound-in-air as our experimental communication with the "real world". (Light, as we will later see, is much trickier to use!)

After years of explaining to "the open minds of students" (rather than pre-programmed colleagues) what is real and what we can measure, I will state my personal belief and viewpoint:

THE ONLY THING ONE CAN ULTIMATELY MEASURE ABSOLUTELY IS THE ACCELERATION OF A PIECE OF MATTER.

Everything else is one or more inferences!

Appealing to Newton's Second Law:

$$a = \frac{F}{m} \qquad (3\text{-}1)$$

where a and m are measured and F is a force which occurs in multiple kinds.

a) m is measured with a reference scale employing Earth Gravity.
b) a is measured via

$$a \equiv \frac{v_f - v_i}{t_f - t_i} = \frac{2d}{t^2} \qquad (3\text{-}2)$$

where the first term is a definition, and the second term is a particular experimentally convenient case with a = constant, v_i = 0, d = distance travelled, t is time elapsed.

For example, in a well-equipped undergraduate laboratory, we might find a force sensor for measuring mechanical (contact) forces. But ultimately, this probe employs a piezoelectric crystal (as does a digital weight scale at home or in your physician's office) that produces an electric current (accelerated electrons) that is proportional linearly to force (over some range of values).

Another example: a charged particle moving through a uniform electric field.

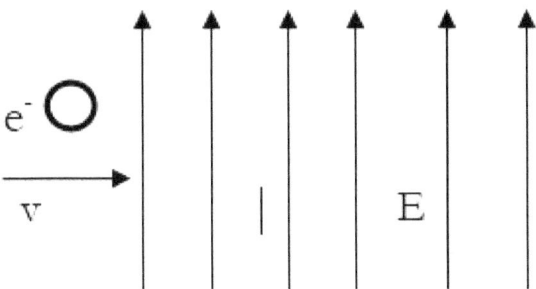

Figure 3-1. The motion of an electron within an Electric Field.

The acceleration of the electron is:

$$a_{along\,v} = 0 \qquad a_{along\,E} = \frac{F}{m} = \frac{-eE}{m}$$

so the acceleration along E depends upon the Electric Field

strength, and the charge and mass of the electron. Standard Newtonian Mechanics (kinematic Mathematics) determine the trajectory. This is a typical problem we give to introductory students.

However, this rendition leads students to reification[28] of Electric Fields that are defined as convenient intermediate quantities.

In fact, this electric field would have been produced by external charges strategically placed, as shown below

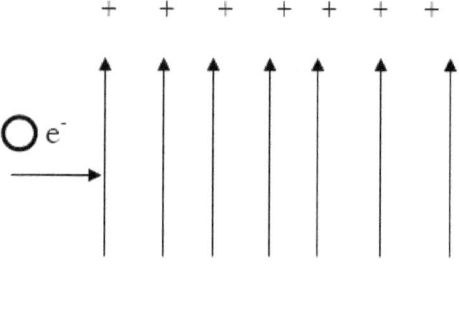

Figure 3-2. The motion of an electron within an Electric Field.

and the actual interaction would have been through the vector sum of the Coulomb's Law force between the electron and <u>all</u> the other positive and negative charges. The vector sum would have been simplified by symmetry (which is a major reason we resort to field theory), and the direct-connection between charges hidden from the student, as well as the nature (still somewhat in question) of the mitigator.

---

[28] "Reification" is the logical error of assuming because one can define something, it must exist.

The "Relative" Is Best Demonstrated By The Doppler Effect For Sound.

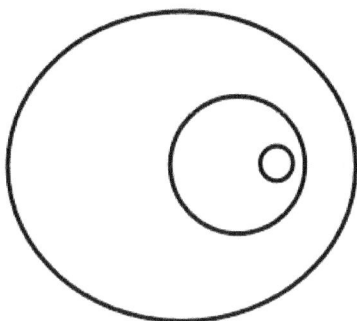

Figure 3-3. Wave fronts at some instant in time created by a sound source moving to the right.

Wave fronts (circles in the diagram) emitted at different times from a source moving to the right are displayed at the same final time. The length between wave fronts is the wavelength, which is inversely proportional to frequency within the medium (air).

$$v_{phase}(\text{air properties}) = 340\, m/s = (frequency)(wavelength)$$

In this situation, the frequency of the source is, say fixed, there is a medium, say air, and there is a receiver. The frequency detected by the receiver is dependent upon *the relative positions and motions of the receiver relative to the source*. Frequencies *perceived* (and measured with instruments) in front of the moving source are increased (wavelength decreased), frequencies perceived behind the moving source are decreased (wavelength increased).

And when a receiver is along a perpendicular line (to the motion of the source, moving with the source), there is *no* frequency shift and the instrument measures the "truth".

As long as one is aware of the disconnect between the receiver's measured frequency and the reality of the situation, one can employ this effect to detect more complicated quantities, such as the speed RELATIVE between the receiver and source (say, you as a motorist and the police).

However, in this setting in particular, a "frequency meter" without extensive context does not measure "absolute truth".

CHAPTER 4: LIGHT, THE BEARER OF QUESTIONABLE TRUTH

Electric Field Subtleties Hidden from Students

Let's revisit *Classical Mechanics*, and the Galilean velocity transformation. If a stream of water (mass) is travelling at velocity u relative to a boat and the boat is moving at velocity w relative to Earth, then the Earth observer measures the water's velocity as u' (as shown below[29])

$$\vec{u'} = \vec{u} + \vec{w} \qquad (4\text{-}1)$$

Next, consider a moving *charged* mass.

Let's do this right. We will follow the staircase steps below[30]:

---

[29] Copyright purchased from https://www.123rf.com/profile_kropic'>kropic / 123RF Stock Photo
[30] NOT the chain in current text books.

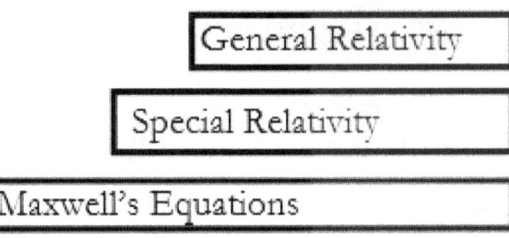

Figure 4-1. The Staircase of Evidence Steps for these Theories.

The coordinate system transformation for the Electric Field of a *moving* charged particle can be derived from the fields of a *stationary* charged particle[31] via Maxwell's Equations (Step 1 of the staircase):

$$\vec{E} = \frac{q}{4\pi\varepsilon_0 r^2}\hat{r} \qquad \vec{E}' = \frac{q}{4\pi\varepsilon_0 R^2}\frac{1}{\gamma}\hat{r} \qquad (4\text{-}2)$$

where $r^2 = x^2 + y^2 + z^2$ and

$$R^2 = \frac{(x-vt)^2}{\gamma^2} + (y^2 + z^2) \text{ and } \gamma = \frac{1}{\sqrt{1 - v^2/c^2}} \qquad (4\text{-}3)$$

A possible sketch of the predicted *observation* of E' might appear as

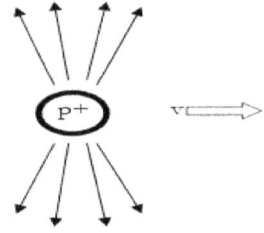

Figure 4-2. The predicted Electric Field (viewed by a stationary observer) of a proton moving to the right.

---

[31] "The Feynman Lectures". Volume 2, Chapter 26. Richard Feynman.

One may interpret this as "*That is what the moving electric field looks like*" or as "*The dimension along the direction of motion has <u>apparently</u> contracted from <u>my</u> perspective*". Stating that the charged particle has contracted because its moving field is contracted is not an obvious conclusion (although many experts categorically claim it IS contracted).

Finally, Coulomb's Law for a stationary charge isn't the whole story. *The electric field doesn't point to where the charge IS* (except if it is stationary relative to the observer and solitary).

Issue 1. For a charge in motion relative to an observer at a field point, the Electric Field is:[32]

$$\vec{E} = \frac{-q}{4\pi\varepsilon_0}\left[\frac{\vec{e}_{r'}}{r'^2} + \frac{r'}{c}\frac{d}{dt}\left(\frac{\vec{e}_{r'}}{r'^2}\right) + \frac{1}{c^2}\frac{d^2}{dt^2}\vec{e}_{r'}\right]$$

where $e_{r'}$ is the unit-direction vector between the charge (at its retarded position r', Eq. 4-4, at time t = r'/c ago). Only for a solitary, stationary charge does the electric field point to the source. Yet this is often the impression left with students.

This electric field also depends upon the first and second time derivatives of properties of the source charge.[33]

Issue 2. The Electric field is the *vector sum* of the electric fields of many source charges:

$$\vec{E} = \Sigma \vec{E}_i$$

In fact, we can see from Gauss' Law that the Electric Field in the far-field only depends upon the net charge inside a surface which includes the field point. Only the near field reveals individual sources.

---

[32] *The Feynman Lectures on Physics*, Vol I, pg. 28-2. Richard Feynman. (1963).
[33] NOT in current text books. Only the stationary charge contribution is initially presented, and then "fixed up" many chapters later without connecting the concepts.

***

It is dangerous to make casual inferences from a measured electric field about where and when the source charge(s) is(are)! This is the same situation as measuring a Doppler frequency. One cannot correctly infer the physical situation without detailed information about the relative location and motion state of the source!

Basis of Second Postulate of Relativity?

Maxwell's Equations (Step 1) thus yield the Lorentz Transformation:

$$x' = \frac{x - vt}{\gamma} \qquad y' = y \qquad z' = z \qquad (4\text{-}4)$$

and this transformation results in the velocity addition expression[34]

$$u' = \frac{u + w}{1 + uw/c^2} \qquad (4\text{-}5)$$

This transformation reveals what we will infer from our instruments about <u>charges and light</u>.

Major point:

If w is itself light, then w = c, and

$$u' = \frac{u + c}{1 + u/c} = c \qquad (4\text{-}6)$$

Or, the velocity of light is measured to be c in any moving frame of reference.

There is no need for the Michelson-Morley result to prove this

---

[34] Griffiths, David J. (2007), Introduction to Electrodynamics, 3rd Edition; Pearson Education – Chapter 12, Problem 12.14.

statement[35], which is Einstein's Second Postulate of Relativity (Step 2 of the staircase).

More Ways Light Can Deceive the Experimentalist

In many instances, light (or any electromagnetic radiation) is the only observing tool we have, especially for rapidly changing events.

But the transmission of information from a moving piece of matter to an observer via light may not be able to keep up with the changes in motion (or state) of the piece of matter. In this case, light WILL NOT correctly measure the behavior of that piece of matter.

We must not fall into the trap of interpreting <u>literally</u> what light overtly conveys to the observer in all cases.

> A fairly common instance of this problem is observing a star at night. In actuality, if the star is far away, it may have burned out long ago and it may take years before the last piece of light emitted reaches the Earth and reveals the change. The star we are "seeing" may no longer be in existence.
>
> Or, if our sun were to burn out, it would be 8+ minutes before we knew it.

Words Can Cause Confusion

A statement commonly made is that light has both wave and particle properties. "Particle properties" is deceiving!

Granted, light is not a continuous wave; it comes in packets (or technically, is quantized into a smallest piece, not a continuum).

---

[35] Which we will refute in the next Chapter.

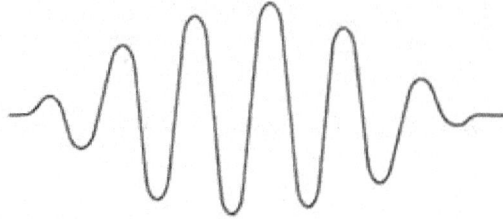

BUT, this packet doesn't travel at any old speed as a particle does. It travels at one speed, $c = 3 \times 10^8$ m/s, and that speed *is the same* in all reference frames that move at constant relative speed. This is a purely wave property, true of light and sound[36].

More Confusion …….

The photon has *zero rest mass*, but the photon is never at rest so this is an oxymoron.

But, the photon has "momentum" which means it has mass when on the move? Really?

Well, not until it interacts with matter. Then the "inertia" or mass is due to the wave's electric and magnetic fields interacting with such matter.

> When our electromagnetic wave hits something *real* (I mean *anything* real, like a wall, or some molecule of gas), it is likely to hit some electron, i.e. an actual electric charge. Hence, the electric and magnetic field should have some impact on it. Now, as we pointed out above, the magnitude of the electric force will be the most important one – by far – and, hence, it's the electric field that will 'drive' that charge and, in that process, give it some velocity *v*, as shown below. In what direction? Don't ask stupid questions: look at the equation. $\mathbf{F}_E = q\mathbf{E}$,

---

[36] Demonstrated later in this book.

so the electric force will have the same direction as **E**.

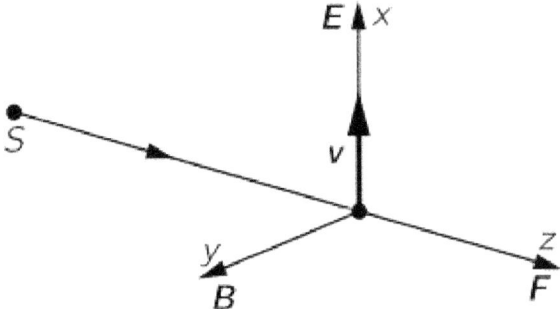

But we've got a *moving* charge now and, therefore, the magnetic force comes into play as well! That force is $\mathbf{F}_M = q(\mathbf{v} \times \mathbf{B})$ and its direction is given by the right-hand rule: it's the **F** above *in the direction of the light beam itself*. Admittedly, it's a tiny force, as its magnitude is $F = qvE/c$ only, but it's there, and it's what causes the so-called radiation pressure (or light pressure *tout court*). – Richard Feynman

Bottom line: The net force transmitted to the surface from the source is (using $J = \sigma E = \rho_c v / A$)

$$\frac{F}{A} \propto (\sigma E) B$$

which is not a true "pressure". Pressure depends only upon parameters of the medium[37]; *this* pressure also depends upon also the <u>conductivity</u> of the surface receiving the force.

---

[37] Gas pressure impinging upon a wall only depends upon the temperature (average kinetic energy per molecule), NOT on the properties of the wall.

The take away,

- This interaction is an action-at-a-distance force acting between light source charges and wall charges. Too much is being loosely ascribed to the mitigating EM fields.
- Imputing a mass (or pressure) to the electromagnetic wave is inappropriate.
- This is NOT to be confused with matter waves, which carry both energy and momentum.

This effect is analogous to the common student misconception that because one can surf (gain forward momentum) on an ocean wave near the beach (boundary), then one can also surf in deep ocean water.

CHAPTER 5: SPECIAL RELATIVITY IN DETAIL

The Michelson-Morley Experiment

The *traditional* pedagogical heart of the derivation of Einstein's Special Relativity is the concluding statement from the Michelson-Morley experiment:

> There is no evidence for a supporting medium for light propagation, the so called *luminous ether*.
>
> Or, the stronger statement, the speed of light is measured to be the same in all inertial reference frames.

However, there are fundamental concerns with employing this experiment as the SR theory basis.

Issues:

A. For this experiment, light transit times are calculated in the reference frame of the ether and the measurement apparatus is in motion. However, it is always safer to perform the calculation in the reference frame of the measurement apparatus. We will begin with calculating there. Then we will recalculate in the other reference frame, adding a missing effect which makes the results of the two calculations agree! They must!

B. There are two forms of "Relativity", one derived from Maxwell's Equations (applied to charged particles, Electromagnetic fields, and light), and the other (*purported* to apply also to uncharged particles) due to the Einstein's Postulates of Relativity.

To avoid these concerns, we will explicitly employ the well-established Step 1 result (Fig. 4-1), Lorentzian velocity addition from Maxwell's Equations:

$$u' = \frac{u+w}{1+uw/c^2} \qquad (5\text{-}1)$$

Note: The frame transformation for light and sound waves between relatively moving source and receiver is the Doppler Effect. The transformation is a function of v, say T(v)[38], which appears reciprocally in the frequency and wavelength:

$$f' = f_0 T(v) \qquad \lambda' = \lambda_0 / T(v)$$

$$c = f_0 \lambda_0 \qquad c' = f'\lambda' = f_0 T(v) \frac{\lambda_0}{T(v)} = f_0 \lambda_0 = c.$$

Wave speed (the speed of an insubstantial point, such as a wave crest) is thus independent of inertial reference frame: a WAVE property. This is NOT a relativity effect!

Why? Generally, any wave phase speed is proportional to the square root of the magnitude of the restoring force divided by the effective inertia. For sound waves, the restoring force magnitude is the gas compression constant B and the inertia is the density ρ. For electromagnetic waves, the restoring force is electric force (magnitude $1/\varepsilon_0$) and the inertia is the magnetic force (magnitude μ_0). In both cases, these factors are frame-transformation independent, and so such transformations MUST NOT appear in the phase velocity.

A. Computing in the rest frame of the apparatus, ether moving to the left:

Path Transit Phases in Detail

Consider any apparatus diagram, such as:
https://commons.wikimedia.org/wiki/File:Michelson-morley.png.

The paths common to both species of light are in red, the unique paths are in green and blue. All L's are equivalent in this reference

---

[38] See good reference for various Doppler transformations:
http://spiff.rit.edu/classes/phys314/lectures/doppler/doppler.html

frame. Light is embedded in the moving ether reference frame and we are observing in the stationary apparatus frame.

Unique paths:

Path A: 45^0 mirror to right mirror

$$t_A = \frac{L}{velocity} = \frac{L}{(c-v)/1-cv/c^2} = L/c \qquad (5\text{-}2)$$

Path B: Right mirror to 45^0 mirror

$$t_B = \frac{L}{velocity} = \frac{L}{(c+v)/1+cv/c^2} = L/c \qquad (5\text{-}4)$$

The total transit time experienced in the direction parallel to the ether motion is

$$t_{parallel} = \frac{2L}{c} \quad t_{parallel} = \frac{2L}{c} \qquad (5\text{-}6)$$

Paths C and D: Transverse to the ether motion

In the perpendicular to ether motion, the light velocity is exactly[39] c and the total phase experienced in the transverse direction is

$$t_{transverse} = \frac{2L}{c} \quad t_{transverse} = \frac{2L}{c} \qquad (5\text{-}7)$$

Of course, this result is merely a restatement of "The speed of light is the constant c in all inertial reference frames."

---

[39] In the apparatus reference frame, the light speed is exactly c; the parallel components due to: (a) the ether flow and (b) aiming into the ether wind velocity (experimental orthogonality) are equal and opposite.

Conclusions:

- There is NO difference in experienced phase between the two directions/paths. These results are EXACT to all orders in v/c.

- This apparatus could NOT have detected the presence of an ether.

B. *Computing in the rest frame of the ether, apparatus moving to the right.*

This is the standard derivation found in textbooks. However, a subtle recently recognized effect was left out: The equal reflection/incident angles rule may not apply to some moving mirrors.[40]

In fact, what was omitted was the contribution to the moving 45^0 mirror.[41]

The Physics of the reflection from the mirror moving relative to the ether was treated as if the mirror were stationary. The oversight was that the wave front *interference model of reflection* provides a more appropriate model (compared to the ray model) when the mirror is *in motion*.

[The calculation of the image current formulation in a mirror is complicated by its orientation and its motion. Therefore, the straight forward wave interference model of reflection will be employed here.]

For a *stationary mirror* at 45^0, the incoming wave front touches the mirror from bottom to top at different times, and using Huygens' Principle emitters, re-emits a wave front in the orthogonal direction as computed by total equal path lengths. See Figure 5-1.

---

[40] A. Gjurchinovski, *Reflection from a moving mirror -- a simple derivation using the photon model of light*, European Journal of Physics **34**, (2012).

[41] P. Marmet, *The overlooked phenomena in the Michelson-Morley experiment*, (before 2005).

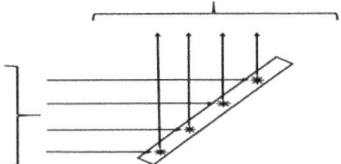

Figure 5-1. For a stationary mirror within the light supporting ether, the interference wave fronts determined by equal path lengths are displayed.

However, when the 45⁰ mirror is *moving* to the right in the ether, the re-emission points are displaced during the reflection process (time progresses from the bottom intersection to the top intersection) and the equal total path lengths (and corresponding transit times) create a wave front that is NOT in the 90⁰ direction. See exaggerated Figure 5-2.

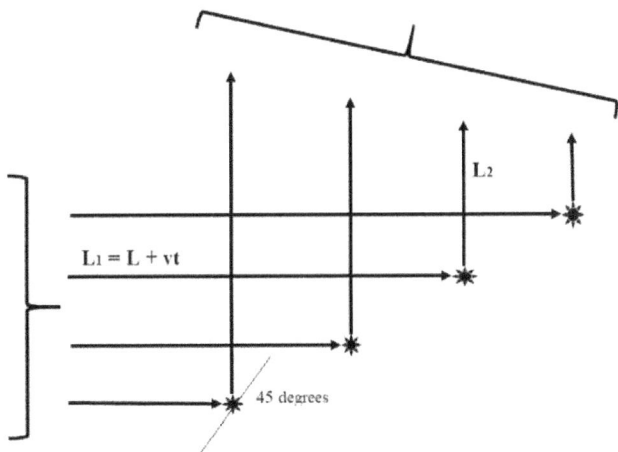

Total Path Length = $L_1 + L_2$ = constant

Figure 5-2. For a mirror moving to the right within the light supporting ether, the interference wave fronts determined by equal path lengths are displayed. The reflected ray does not obey the equal incident/reflected angle rule.

The result is that the light ray (perpendicular to the wave fronts) is tilted forward by and angle α where $\cos\alpha = v/c$. For reference, the Earth's orbital velocity about the sun is about $v/c \approx 10^{-4}$.

When the light is reflected from the transverse mirror (where the equal angle law probably *is* true), it reflects forward by the same angle.

Note: *This is why an observer moving with the apparatus at velocity v can observe the ray traveling <u>exactly</u> orthogonally relative to that observer.*

This lengthens the transit times in the <u>transverse</u> direction!

<u>Path Transit Times in Detail</u>

Path 1: 45^0 mirror to right mirror

The experienced path length is increased[42]

$$ct_1 = L + vt_1 \qquad (5\text{-}8)$$

because the right mirror is moving to the right during transit and

$$t_1 = \frac{L}{c-v} \qquad (5\text{-}9)$$

Path 2: Right mirror to 45^0 mirror

The experienced path length is decreased

$$ct_2 = L - vt_2 \qquad (5\text{-}10)$$

because the 45^0 mirror is moving to the right during transit and

$$t_2 = \frac{L}{c+v} \qquad (5\text{-}11)$$

The phase accumulated during paths 1 and 2 becomes:

---

[42] The way this calculation as performed in standard texts looks like Galilean transformation was employed for light, which cannot be correct. Here, we will follow Feynman's clear and correct analysis.

$$\phi_{parallel} = \frac{\omega L}{c}(\frac{1}{1+v/c}+\frac{1}{1-v/c}) = \frac{\omega L}{c}\frac{2}{1-v^2/c^2} \simeq \frac{2\omega L}{c}(1+v^2/c^2) \quad (5\text{-}12)$$

Paths 3 and 4: Transverse light rays

In the ether frame of reference, the light path is an isosceles triangle because of the movement of the apparatus to the right, with an apex angle θ, where cosθ = v/c. See Figure 5-3.

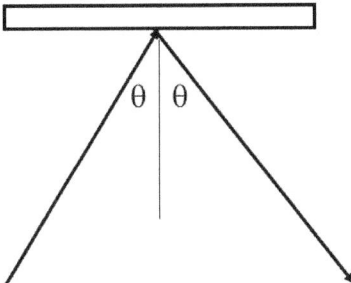

Figure 5-3. The actual diagonal paths of the light rays due to: a) motion of the apparatus through the ether, and b) the altered reflection angle from the moving mirror.

This results in a path length (and transit time) lengthened by

$$\sqrt{1+v^2/c^2}$$

We also must account for the above mentioned lengthening due to the transmission angle α, where cosα = v/c. See Figure 5-3 with α = θ.

This also results in a path length (and transit time) lengthened by

$$\sqrt{1+v^2/c^2}$$

The total path length then becomes:

$$L_a = L\sqrt{1+v^2/c^2}\sqrt{1+v^2/c^2} = L(1+v^2/c^2)$$

$$t_a = \frac{L_a}{c} = \frac{L}{c}(1+v^2/c^2) \tag{5-13}$$

and there are two of these.

The accumulated phase then becomes:

$$\phi_{transverse} = \frac{2\omega L}{c}(1+v^2/c^2) \tag{5-14}$$

Strictly speaking, the wave front must be traced through the entire system until it is parallel to the wave front from the left-right light path. That is shown in Figure 5-4.

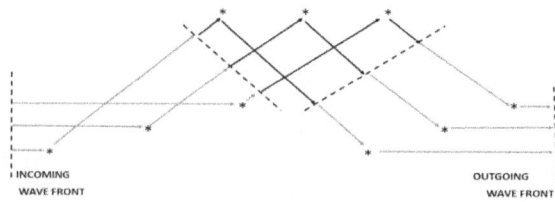

Figure 5-4. The wave front history of the transverse light path until it becomes a wave front that can interfere with the left-right path.

Conclusions:

There is NO difference in phase (equations 5-12 and 5-14) to second order between paths in the ether frame of reference.

This null result alone does not mean that the Second Postulate of Relativity nor absence of ether are wrong. It just means the Michelson-Morley experiment didn't prove them!

What about the Singularity?

As seen in a previous Chapter, Feynman showed that the Lorentz

Field transformation contains the γ factor. See Eq. 4-3.

However, there is a singularity in the Electric Field strength if the charge's velocity equals the speed of light. The energy required to achieve this state approaches infinity.

But, that makes sense.

At the speed of light, the electric field cannot exist/be emitted to the right of the charge at all, and the total electric field is collapsed into an infinite two-dimensional pancake of zero width.

Surely, the finite size of the electron must come into play and somehow limit the contraction. The point-charge approximation cannot be true for this level of contraction.

Regularization needs to be considered[43]

> In physics, especially quantum field theory, **regularization** is a method of modifying observables [measured quantities] which have singularities in order to make them finite by the introduction of a suitable parameter called **regulator**. The regulator, also known as a "cutoff", models our lack of knowledge about physics at unobserved scales (e.g. scales of small size, or large energy levels). It compensates for (and requires) the possibility that "new physics" may be discovered at those scales which the present theory is unable to model, while enabling the current theory to give accurate predictions as an "effective theory" within its intended scale of use.

In fact, Feynman proposed several models of a charged particle to

---

[43] Wikipedia.

resolve this issue.[44] Feynman made NO claim that charged particles could not travel up to or faster than light based upon this singularity.

More Ways Light Deceives

Special Relativity involves the predictions of *observations* made by observers moving at constant velocity relative to each other.

Issue 1: Validity of inferences made about events in a moving frame

One cannot validate what one observes in a coordinate system moving relative to that observer without:

- A. Sending "information" between observers during and after an event via Electromagnetic radiation, which we have shown is risky
- B. Decelerating the moving frame, reversing the motion (accelerating) and comparing when the observers are in the same place at the same time. This is the subject of General Relativity (GR) which we won't address here, but IS experimentally well verified. Unfortunately, most "experimental verifications" of SR are GR results.

So what a stationary observer "measures" about the events in a moving coordinate system is subject to interpretation, not necessarily an *absolute* connection to reality for the moving observer.

The moving observer, however, must make Physics sense of their observations, see Relativity Postulate 1 in Chapter 2.

Issue 2: Simultaneity

Two events that are simultaneous (at the same instant in time) to one observer may be at different times in a moving coordinate system.

---

[44] "The Feynman Lectures". Volume 2, Chapter 28. Richard Feynman.

Consider the situation below. To avoid concept confusion, let us use *sound* as the communication mechanism for measuring the simultaneous lightning strikes (for the observer on the ground):

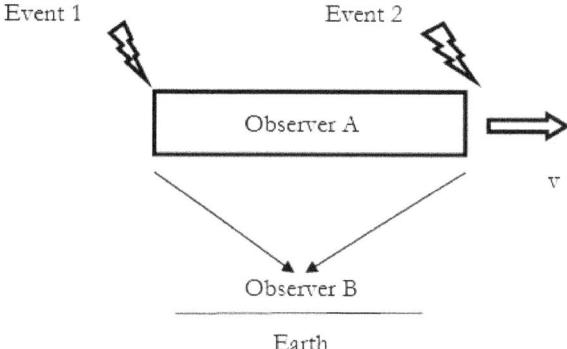

Figure 5-6. Lightning strikes the ends of a trolley simultaneously, according to what Observer B hears. What does Observer A hear?

To the ground observer, the sound from the two lightning strikes (events) arrives at the observer at the same time. For the on-board observer moving to the right, the sound from event 2 reaches that observer before the sound from event 1 because of the motion. So one infers that the on-board observer will perceive event 2 to occur in time before event 1. And in this case, the inference is accurate as to *what the moving observer will hear*. But this result is to be expected for waves, similar to the Doppler Effect, where relative location and motion of the source and observer influence perception and measurement.

But how does the *perception* of the on-board observer effect the *reality* of the simultaneous lightning strikes on the ground? Not at all.

The trolley traveler can understand the Earth observer's interpretation if she looks out the window and notices her movement.

This is akin to observing events while in an accelerated frame of reference.

See animation at:

https://en.wikipedia.org/wiki/Coriolis_force

One has to invoke "fictitious forces" for an observer in an accelerating observation frame to reconcile observations with the expectation of observing in a non-accelerating frame.

For the sake of simplicity and clarity, one just should avoid observing in an accelerated frame unless one is FULLY aware of the implications for those observations.

Issue 3: What Is Real and What Is for Calculation Convenience?

Let's consider an empty universe except for two pieces of matter, which have mass and charge. Each will begin falling toward the other, experiencing what we call action-at-a-distance forces:

$$F_{gravity} = G \frac{m_1 m_2}{d^2} \quad F_{electric} = \frac{1}{4\pi\varepsilon_0} \frac{q_1 q_2}{d^2} \tag{5-15}$$

where m's and q's are the respective masses and charges, and d is the center-center distance between the pieces of matter. That means that via some mechanism across possibly immense distances, the values m_1, m_2, q_1, q_2 and d must be transmitted between masses and charges. No one knows how this is accomplished.

There are currently three candidates for this "mitigator" of forces

1. Fields
2. Deformations into an unknown dimension, known as potentials
3. Exchange of virtual particles.

Mathematically, the first two choices are represented in space surrounding the masses as a spatial three-dimensional number-mesh, one vector and one scalar (related by a vector differential operator), and provide exact predictions of behavior which are the basis of engineering devices. No one knows which or if either is the "real mitigator".

The third choice provides little quantitative predictive ability. But the mitigating virtual particles (photons, zero mass) for electricity (and multiple mitigators for the nuclear force) have been experimentally isolated, while the postulated mitigator for gravity (gravitons) have not.

The so-called "Theories of Everything" are an attempt to find a single paradigm or explanation for all forces.

Meanwhile, fields provide an engineering design basis.

The danger is in believing that this mitigator (an EM field) is *real* rather than a calculational mental-construct with accurate predictability. As seen in the example of the Electric Field previously, there is not a one-one correspondence between electric field and sources.

Electromagnetic radiation is a slippery tool when used to reveal experimental truth.

1. Light that we capture provides ONLY information about:

    - The direction from which it last encountered matter (mirrors and lenses).

2. Light so captured does NOT provide information about:

    - Where this light originated

    - How long it has been in flight.

3. However, light is USEFUL and of experimental expedience/necessity when:

- Observing stars

- Observing scattering from matter too small to be observed by other means

- Observing matter moving near the speed of light.

Usefulness, as delineated in the above list, however, has often led experimenters to forget the limitations of light and to interpret captured light information *at face value*.

Additional Issues

Location of the emitter

When a charge moves suddenly from one spatial point to another (accelerates), the information that this event has occurred does not move instantaneously throughout space. The information moves outward in a sphere of radius

$$R = c\Delta t$$

and the electric field inside that moving sphere points to where the charge "is", while the electric field outside the moving sphere points to where the charge "was".

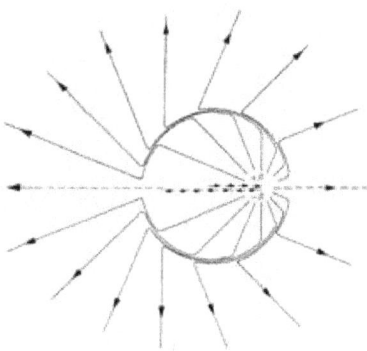

Figure 5-7. The Electric Field of an accelerated charge at some later instant in time.

Time delays

If a light ray encounters a piece of matter in transit, Huygens's Principle (which is true based upon Atomic Physics) states that the original light before encounter is absorbed, held for a delay time, and then *new* light is emitted. That time delay may depend upon the motion state of the piece of matter involved.

Transit times

All too often, equations of motion for *matter* are employed with *light rays* $\Delta t = \dfrac{L}{v}$ and v is replace by c cavalierly.

As shown above, if there is motion of a reference frame involved, the speed of light in all reference frames is an absolute c, and may obscure the transit time severely.

Einsteinian Relativity

Time Dilation

Consider Einstein's "light clock". A light pulse is sent out perpendicular to the motion of a vehicle, reflected, and two observers measure the times of transit in both coordinate systems, one ground-based and one vehicle-based:

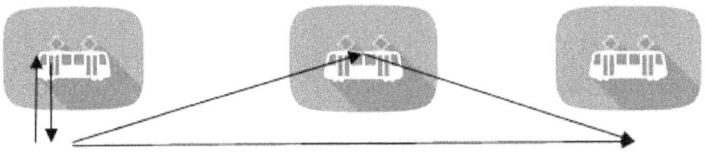

Figure 5-8. Einstein's "Light Clock" as reported by an Earth-based observer and a trolley-based observer.[45]

---

[45] Copyright for Trolley purchased from htpp://www.123rf.com.

$$t_{Earth} = \frac{2D}{c}$$

$$t_{Vehicle} = \frac{2}{c}\sqrt{D^2 + (vt_{vehicle})^2}$$

$$t_{vehicle} = \frac{2D}{c} \frac{1}{\sqrt{1-v^2/c^2}} \quad (5\text{-}19)$$

Thus, the Earth observer may *infer* that the "clock" of the moving events is dilated by a velocity-dependent amount.

So what?

Did the moving system *really* have the system clock altered by the motion, or is this just the *perception* of the Earth observer? Actually, this *perception* may be just the intrinsic lag in information propagation noted above (lack of simultaneity). One does NOT have to interpret this as a *real* change in the clock in motion.

Again, knowing the limitations of light regarding one-to-one communicate of information, this form of "light clock" should not be taken seriously.

Note: In General Relativity ($a \neq 0$), there IS an actual change in the accelerated clock. See arguments below. Do not confuse!

Length Contraction

Note: The derivation of length contraction DEPENDS upon the derivation of time dilation; that brings forward the caveats from time to length!

Consider the same trolley and observer, and a rod of length L_p lying along the direction of travel, where the "proper" label refers to the reference frame in which the measurement instrumentation lies. The trolley dweller sends a laser pulse along the rod from one end, the light reflects from a mirror at the other end, and returns to the light receiver co-located with the laser. The time interval for send (event 1) to receive (event 2) is

$$\Delta t_p = \frac{2L_p}{c} \quad (5\text{-}20)$$

The Earthbound observer also records events 1 and 2. The time interval measured in the Earthbound reference frame is

$$\Delta t = \frac{2L}{c} \quad (5\text{-}21)$$

where L is the *imputed* length in the Earth frame.

Thus, dividing these two equations,

$$\frac{L}{L_p} = \frac{\Delta t}{\Delta t_p} = \sqrt{1 - \frac{v^2}{c^2}} \quad (5\text{-}22)$$

or,

$$L = L_p \sqrt{1 - \frac{v^2}{c^2}} \quad (5\text{-}23)$$

The length in the Earth frame is computed to be shorter than the moving rod length.

Well, that is cute, and it is a standard derivation in text books.

BUT, it ASSUMES the clock in frame O has changed, and ignores the alternative explanation that this is merely asynchronicity.

Problem? Time asynchrony (Eq. 5-19) is introduced by a light beam moving TRANSVERSE to the motion (i.e., extra path length), and is not *clearly* applicable here.

Let's go at this a different way, and deal with the light moving parallel to the motion correctly.

Because of the <u>moving mirror</u> at the far end of the rod, the path traveled by light downstream is longer (mirror moving away from light) and the upstream path shorter (detector moving toward light).

Notes: We saw this in the Michelson-Morley apparatus. And what follows mathematically *looks like* light traveling slower and faster than the quoted constant speed c, but the ultimate cause of the mathematics is as described above.

$$\Delta t_p = \frac{2L_p}{c}$$

$$\Delta t = \frac{L}{c-v} + \frac{L}{c+v} = \frac{2L}{c} \frac{1}{1-\frac{v^2}{c^2}} \quad (5\text{-}25)$$

a larger asynchrony than in the transverse direction. So a single sliding-scale clock won't work.

Well, it will if one allows length contraction along the direction of motion

$$L = L_p \sqrt{1 - \frac{v^2}{c^2}} \quad (5\text{-}26)$$

So, either these time differences are lack of simultaneity (and are anisotropic), or clocks and (parallel) lengths vary with inertial observer.[46]

It isn't obvious to me which is correct, but the community of Physics has chosen to reify both distortions in length and time.

Many of my colleagues still say these dilations and contractions are *appearances from the Lab frame perspective*, not actual changes.

---

[46] Figure 4-2 should provide some credence to length contraction, but I am still wary of that interpretation.

CHAPTER 6 DARK MATTER

The Historical Setting

When the Hubble telescope began transmitting back volumes of new Cosmological Data and Photos, a peculiarity in the dynamics (motion) of galaxies became apparent. A typical Spiral Arm galaxy might look like:[47]

Figure 6-1. A spiral arm galaxy.

Note that there is a central bulge of stars of one *general* type of star, and an outer disc of different *general* type of star.

The expectation for the tangential velocity of orbiting stars was that this velocity would first increase radially *within* the bulge as the mass within the orbit increased, and then drop off (Kepplerian dynamics, such as in our planetary system) in the mass-sparse disc where stars would be relatively decoupled from each other as our solar system planets are. When transverse galactic arm orbital velocities were mapped, however, the results were surprising: the outer disc stars

---

[47] Copyright purchased from http://www.123rf.com/profile_alexmit'>alexmit / 123RF Stock Photo.

rotated at same tangential velocity.

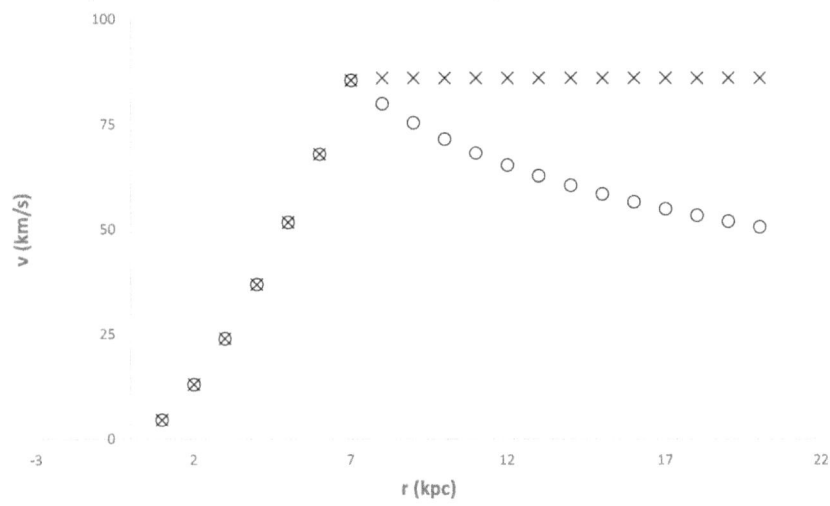

Figure 6-2. Representation of data: The star orbital velocity versus radius from the center of the galaxy. The circles are what would be expected for Kepplerian motion; the x's are what is found experimentally.

Interpreting these Results

1. Notice that the velocities increase until all the bulge mass is encircled according to Kepler's Law. These stars are moving as individual, non-interacting masses, responding to the gravity of the stars below their orbits. Gauss' Law for gravity.

2. In the disc region, the velocity profile is surprisingly flat. These stars seem to behave collectively in response to something else. There is "new Physics" in the disc.

Dynamics Background

To travel in a closed orbit in the outer disc, you need an inward acceleration

$$a_{centripetal} = \frac{v_{tan\,gential}^2}{r} \tag{6-1}$$

and that acceleration is provided by gravity

$$a_{gravity} = G\frac{m_{inside-orbit}}{r^2} \tag{6-2}$$

Hence, orbital velocity becomes

$$v_{tan\,gential}^2 = G\frac{m_{inside-orbit}}{r} \tag{6-3}$$

In a classical description, for this velocity *in the disc* to be constant (see Figure above), the mass in the disc needs to increase like r.

$$m_{inside-orbit} = m_0 r \tag{6-4}$$

so as to remove any r-dependence in the velocity equation above.

Yet optical observations do not support that mass increase.

The Astronomy fix: Invent mass that does not emit light (regular matter does unless at absolute zero which the Third Law of Thermodynamics says it can't attain). This was a logical way to proceed 50 years ago, since the paradigm *at that time* was that Newton's Universal Gravity formulation *alone* was sufficient to explain the workings of the universe, and mass was the free parameter.

Today, thanks to the Hubble Telescope, we have exponentially more information about the structure of the universe and other candidate parameters at our disposal.

The dark matter approach, unfortunately, *forces the required behavior* rather than *discovers and explains the cause* of the *observed* behavior. See more concerning this criticism in an alternative formulation.[48]

I might also add that as time went on, the community piled on properties to dark matter each time they needed to explain some other anomaly. Dark Matter became a catch-all explanation for any new unexplained phenomenon.

In addition, investigators in this area have clung, until today, to dark matter as THE explanation because they WANT the density of matter in the universe to attain the critical value such that the universe doesn't end.[49] There is not sufficient visible matter to attain this want.

[The consequence of having less than this critical density is that gravity is insufficient to stop the universe from ongoing expansion, old stars dying become too far away from new stars forming to seed them, no new stars form, old stars go to final dead state, and the universe becomes disperse and dark.]

I agree with Erik Lerner[22] that Big-Bangers want there to be a purposeful beginning to the universe. And I also claim they have a need to have a meaningful sustained final universe rather than a purposeless decay into nothingness.

Religion and Science have again become entangled.

My Perspective as a Physicist

Let me look at this as a Physicist rather than as an Astronomer.

Physicists only deal with concepts they can model and measure. So hypotheticals, like dark matter and dark energy, which *by definition* cannot be measured, are philosophically and procedurally unacceptable.

---

[48] "The Big Bang Never Happened". Eric J. Lerner, 1991.
[49] Critical density is the value at which the Universe is at balance, and expansion is stopped. -- Wikipedia.

I Claim "missing mass" (if it exists) means that mass has indeed been conserved, but other-represented as some form of energy.

Example

In Chemistry, if you add the weight of an electron and a proton (the components of atomic hydrogen) you will find that the atomic weight for atomic hydrogen is less than simple addition of parts. Indeed, atomic hydrogen is electrostatically bound, that bond contains energy, and that energy is accounted for by missing mass.

$$\Delta m = \frac{E}{c^2} \qquad (6\text{-}5)$$

Also, in nuclear physics, for both fission and fusion processes, the bound nucleus weighs less than the components. Again, the mass has decreased and reappeared as energy in the nuclear bond.

- *I would interpret any "missing mass" as a form of energy, NOT actual mass.*

A better approach would be to multiply the "missing mass" in rotation curves by c^2, and look for an energy source of that magnitude.

An obvious candidate would be the galactic binding energy, but that is too small.

The search continues, but now, with this approach, one is looking for an energy source within classical Physics of a certain magnitude. This approach is TESTABLE. Not buried within a construct that cannot in-principle be tested.

- *In the disc region, it is assumed that gravity is the only acting force, yet current knowledge does not support that. I would look for a "missing force".*

Again, since the dark matter proposal, the Hubble telescope data has grown exponentially what we know about the Physics of the universe.

Why give up on known Physics explanations?

So, suppose some classical Physics is possibly at work! Let's look for it.

What Kind of Mechanism Might We Look For?

<u>Preliminary Issue: Stellar Nuclear Fusion Dynamics</u>

Dependent upon the original mass of matter that contracted to form the star, the internal temperature of a compressed star would allow increasingly higher energy final nuclear fusion products.

For use in later arguments, I would like to classify these fusion reactions by their interaction with the environment: Input and Output products.

Consider the following chain of fusion reactions

Note: Hydrogen nuclei are protons.

Category 1 Stars: Proton-Proton Burners

$$6 \text{ protons} \rightarrow He + 2 \text{ neutrinos} + 2 \text{ protons}|_{ejected}$$

Each proton is ejected with 2.3 Mev of energy. These stars also produce Helium in their end-stage of development. After all the hydrogen is converted to Helium, the nuclear burn ceases, the star collapses further until helium-helium burning occurs. But, when the next level of fusion burning begins, there is an expulsion of Helium beyond a remaining to-be-burned core of Helium. These are the red giant stars. Free Helium is released.

Category 2 Stars: Heavier Element Burners

$$^{12}C + 1 \text{ proton}|_{absorbed} \rightarrow {}^{13}N + \text{photon}$$

$$^{13}N \rightarrow {}^{13}C + \text{positron} + \text{neutrino}$$

$^{13}C + 1\ proton\,|_{absorbed} \rightarrow\ ^{14}N + photon$

$^{14}N + 1\ proton\,|_{absorbed} \rightarrow\ ^{15}O + photon$

$^{16}O + ^{4}He\,|_{absorbed} \rightarrow\ ^{20}Ne$

$^{20}Ne + ^{4}He\,|_{absorbed} \rightarrow\ ^{24}Mg$

$^{28}Si + 7^{4}He \rightarrow\ ^{56}Fe$

Category Description

Category 1 reactions occur in smaller suns that burn until the hydrogen content is used up. Of course, charge is conserved, and if we had kept track of electrons, two of these would be free also, unbound forming a Hydrogen plasma.

The smaller suns that process this reaction do so slowly and live long.

Category 2 reactions occur in larger stars and are located mostly in the galactic disc (we determine that by metallicity; heavier stars fuse elements up to iron). These are proton and Helium absorbing reactions. These larger mass stars burn quickly, and are new.

My Theory: The Change in Physics within the Disc

Suppose that gravity due to the galactic bulge mass is NOT the only force involved in disc dynamics.

Let's make a model. Within the disc, <u>suppose</u> the net force changes in radius to the first inverse power

$$\frac{mv_{tan\,gential}^{2}}{r} = \frac{mC}{r} \qquad (6\text{-}6)$$

And suppose the mass of a disk star is proportional to its cross-sectional area, A. Then:

$$\frac{mv_{\text{tangential}}^2}{r} = \frac{AC'}{r} = Ap \qquad (6\text{-}7)$$

and the force we are looking for exists as a pressure

$$p = \frac{C'}{r} \qquad (6\text{-}8)$$

This is new but classical Physics at work! Suppose, for instance, that there is a fluid flowing inward along the disc with sufficient speed to cause effects stronger that gravity.

Then as the fluid flows inward, its pressure increases like $\frac{1}{r}$ (a one dimensional compression), matching the assumptions above.

Figure 6-4. The increase in the pressure in the galactic disc is a function of radius.[50]

---

[50] Copyright purchased from http://www.123rf.com/profile_sudok1'>sudok1 / 123RF Stock Photo

Again, the important point is that looking at the problem from this concrete view, and knowing the *form* of the unknown force from galactic rotation curves, these ideas are TESTABLE!

What is this "New Force"?

There are a several candidates.

1. If galaxies contain black holes, then the influx of accretion mass could be this fluid flow, and it could be large.

2. The flow of Hydrogen and Helium from Category 1 stars to Category 2 stars. To model this, one would need to know the exact densities and flow rates of these Category stars in both the disc and bulge.

   For NGC 6946 such hydrogen gas flow has been detected.[51]

   > Inpouring rivers of hydrogen gas could explain how spiral galaxies maintain the constant star formation that dominates their hearts, a new study reports.
   >
   > Using the Green Bank Telescope (GBT) in West Virginia, scientists observed a tenuous filament of gas streaming into the galaxy NGC 6946.... The find may provide insight into the source of fuel that powers the ongoing birth of young stars, researchers said.
   >
   > "We knew that the fuel for star formation had to come from somewhere," study lead author D. J. Pisano, of West Virginia University, said in a statement. "So far, however, we've

---

[51] D. J. Pisano (WVU); B. Saxton (NRAO/AUI/NSF); Palomar Observatory – Space Telescope Science Institute 2nd Digital Sky Survey (Caltech); Westerbork Synthesis Radio Telescope. From http://www.space.com/24780-spiral-galaxies-hydrogen-gas-river.html.

detected only about 10 percent of what would be necessary to explain what we observe in many galaxies."

3. There is an alternative cosmology theory proposed by Nobel Laureate Hannes Alfven[52] and a student Erik Lerner, which is now becoming more legitimate as the predicted features (magnetic fields for one) are being discovered. The assertion of this theory is that electro-magnetic force is also active and much stronger and more complicated/structured than gravity. Gravity is a purely radial, smoothing force and too weak. (Dark Matter was introduced to "beef it up".) However, electro-magnetic forces are larger than gravity and have azimuthal (angular) components that can produce many of the complicated structures we see in the universe that are not possible with gravity alone.

Indeed, scale models of electro-magnetic galaxies have been created in an Earth-based laboratory.[53]

And, there is a recent map of the hydrogen distribution within the Milky Way that should provide the sufficient detail of this atomic Hydrogen flow.[54]

A map of helium concentrations and flow would also be useful.

4. The galactic magnetic fields (and associated force) *within the disc* may well behave as 1/r.

> **Magnetic fields** are a major agent in the interstellar medium (ISM) of spiral, barred, irregular and dwarf galaxies. They contribute significantly to the total

---

[52] Alfven published almost exclusively in IEEE (THE electrical engineer's journal), which wasn't read by other disciplines.
[53] "The Big Bang Never Happened". Eric J. Lerner, 1991. Maxwell Laboratory experiments.
[54] "This Is the Most Detailed Hydrogen Map of the Milky Way Galaxy Ever". George Dvorsky, submitted to *Astronomy* October 2016.

pressure which balances the ISM against gravity. They may affect the gas flows in spiral arms, around bars and in galaxy halos. Magnetic fields are essential for the onset of star formation as they enable the removal of angular momentum from the proto-stellar cloud during its collapse. Magneto-hydrodynamic (MHD) turbulence distributes energy from supernova explosions within the ISM. Magnetic reconnection is a possible heating source for the ISM and halo gas. Magnetic fields also control the density and distribution of cosmic rays in the ISM.[55]

One description of the role of these magnetic fields is discussed in a reference.[56] The comment is that although these fields are small in magnitude, they are able to create large pressures within the plasma components found ubiquitously within galaxies and their outlying structures.

5. Finally, there is a proposal that has been repeatedly suggested and then fallen into disrepute and revived again. The form of Newton's Law of Universal Gravitation

$$|F| = G\frac{m_1 m_2}{r^2} \qquad (6\text{-}9)$$

may be a local approximation to the true Universal Law, which might consist of an inverse power series. That is, at smaller cosmological scales, the inverse-square law applies, and at larger cosmological scales the inverse first-power law is dominant. This is sometimes referred to as "Modified Newtonian Dynamics"[57] or more recently as the "Dark Matter Conspiracy".

---

[55] Rainer Beck (2007), Scholarpedia, 2(8):2411.
[56] https://astrobites.org/2015/09/03/a-magnetized-universe-how-galaxies-are-influenced-by-magnetic-fields/
[57] Milgrom, M. (1983). "A modification of the Newtonian dynamics as a possible alternative to the hidden mass hypothesis". *Astrophysical Journal.* **270**: 365–370.

CHAPTER 7: CONCLUSIONS

I have outlined several Physics and Astronomy dogmas that seem to be failing.

Despite tradition and resistance to change, these ideas need to be revisited with an open mind. Even additional calculations[58] and opinions of Nobel Laureates have been conveniently forgotten or steamrolled!

Concerns:

1. As of the 1990's, too many specialty Journals (and sub-disciplines) emerged, and one specialty (or discipline) stopped reading other specialties. Important new results were not updated in introductory Physics texts, which is where Astronomers learn most of the Physics underlying the theories they develop.

2. It takes *at least* 50 years of careful scrutiny to hunt down and dispatch all possible error sources and alternative explanations before any concept can really be considered reliably vetted.

This is the scientific process, and we will always only know a MODEL of the Truth. Neither the general public nor researchers themselves should get overly-invested in, nor overly-defend, the *current* understanding.

3. Exciting "new" results are often appearing in introductory text books as vetted truth. This is a dangerous practice! It leads to lack of critical *continuous* thinking about Physics concepts, since they are relegated to student's sub consciousness early in their career (when they are less critical and more brain-strained) and then tend to remain in the subconscious as "given".

---

[58] In addition to the work for which the Prize was awarded.

www.ingramcontent.com/pod-product-compliance
Lightning Source LLC
Chambersburg PA
CBHW061217180526
45170CB00003B/1045